FACT FINDERS
KNOW THE DIFFERENCE!

I0807260

INSECT OR ARACHNID?

EZRA E. KNOPP

PowerKiDS press™

Published in 2026 by The Rosen Publishing Group, Inc.
2544 Clinton Street, Buffalo, NY 14224

First Edition

Editor: Greg Roza
Book Design: Michael Flynn

Photo Credits: Cover (ant) Wirestock Creators/Shutterstock.com; cover (scorpion) DWI YULIANTO/Shutterstock.com; p. 4 Mezee Ji/Shutterstock.com; p. 5 (beetle) Marwannka30/Shutterstock.com; p. 5 (spider) woravit thongpolyos/Shutterstock.com; p. 6 Paulrommer SL/Shutterstock.com; p. 7 Wanchai Orsuk/Shutterstock.com; p. 8 https://commons.wikimedia.org/wiki/File:Wetapunga.jpg; p. 9 PetlinDmitry/Shutterstock.com; p. 10 David Havel/Shutterstock.com; p. 11 will_dupratt/Shutterstock.com; p. 12 chinahbzyg/Shutterstock.com; p. 13 Catherine Eckert/Shutterstock.com; p. 14 Lauren Suryanata/Shutterstock.com; p. 15 Gulf MG/Shutterstock.com; p. 16 nico99/Shutterstock.com; p. 17 Pierre Williot/Shutterstock.com; p. 18 YingHui Liu/Shutterstock.com; p. 19 Yevhen Prozhyrko/Shutterstock.com; p. 21 retsel/Shutterstock.com.

Cataloging-in-Publication Data

Names: Knopp, Ezra E.
Title: Insect or arachnid? / Ezra E. Knopp.
Description: Buffalo, New York : PowerKids Press, 2026. | Series: Fact finders: know the difference! | Includes glossary and index.
Identifiers: ISBN 9781499452723 (pbk.) | ISBN 9781499452730 (library bound) | ISBN 9781499452747 (ebook)
Subjects: LCSH: Insects–Juvenile literature. | Arachnida–Juvenile literature.
Classification: LCC QL467.K56 2026 | DDC 595.7-dc23

Manufactured in the United States of America

CONTENTS

CREEPY CRITTERS

How do you feel around bugs? Although they may be creepy to some, bugs are a fascinating topic to study! In fact, scientists are still gaining knowledge by studying bugs and all the creepy things they do. But...what is a "bug"?

Bug is a term often used to describe two animals: insects and arachnids. They share things in common, such as jointed legs and a lack of a backbone. But how can you tell them apart? Let's **explore** the world of bugs to find out!

CAN YOU TELL WHICH OF THESE BUGS IS AN INSECT AND WHICH IS AN ARACHNID?

LOOK AT THE LEGS

Insects and arachnids may seem similar on first look, especially when they are very small. But there's one easy way to tell them apart. Count the legs! Which has six legs? Which has eight legs?

Ants are insects, and all insects have six legs. Spiders are arachnids, which have eight legs. You might be wondering, what about centipedes and millipedes—they have tons of legs! Centipedes and millipedes aren't insects or arachnids. However, all three of them are **arthropods**.

BIG BUGS!

Many insects and arachnids are tiny. That's sometimes why people find them creepy—they can sneak up on you! However, not all are tiny. In fact, some are huge! Which do you think has the heaviest species, insects or arachnids?

GIANT WETA

The giant weta insect of New Zealand looks like a huge grasshopper. Their long, thick bodies can weigh up to 2.5 ounces (71 g). However, the Goliath birdeater tarantula of South America has the weta beat. These huge spiders can weigh up to 6 ounces (170 g)!

WILD WINGSPANS

When it comes to flying, insects have the right body parts, while arachnids don't. We should be glad there are no flying spiders! Fleas almost seem like they can fly, but they're just super jumpers. So if insects can fly, which species have the biggest wings?

While there are several moths that come close, the Atlas moth has the largest insect wings by area—close to 25 square inches (161 sq cm)! The **wingspan** of the Atlas moth is about 9.4 inches (23.9 cm).

TINY BUGS

We've observed some of the largest insects and arachnids, but some of them are so small they're hard to see without a **magnifying lens**! Which group has the smallest creature, insects or arachnid?

Fairyflies are really small. The smallest fairyflies are only about 0.02 inch (0.5 mm) long. However, arachnids called mites are even smaller. Some can only be seen clearly with a **microscope**. The smallest mites are only about 0.004 inch (0.1 mm) long!

DEADLY!

Some insects and arachnids aren't just creepy, they're deadly! The tiny fire ant **injects** a painful **venom** with its bite that causes great pain. The deathstalker scorpion is the most venomous scorpion in the world. However, there's one clear winner of deadliest insect in the world.

DEATHSTALKER SCORPION

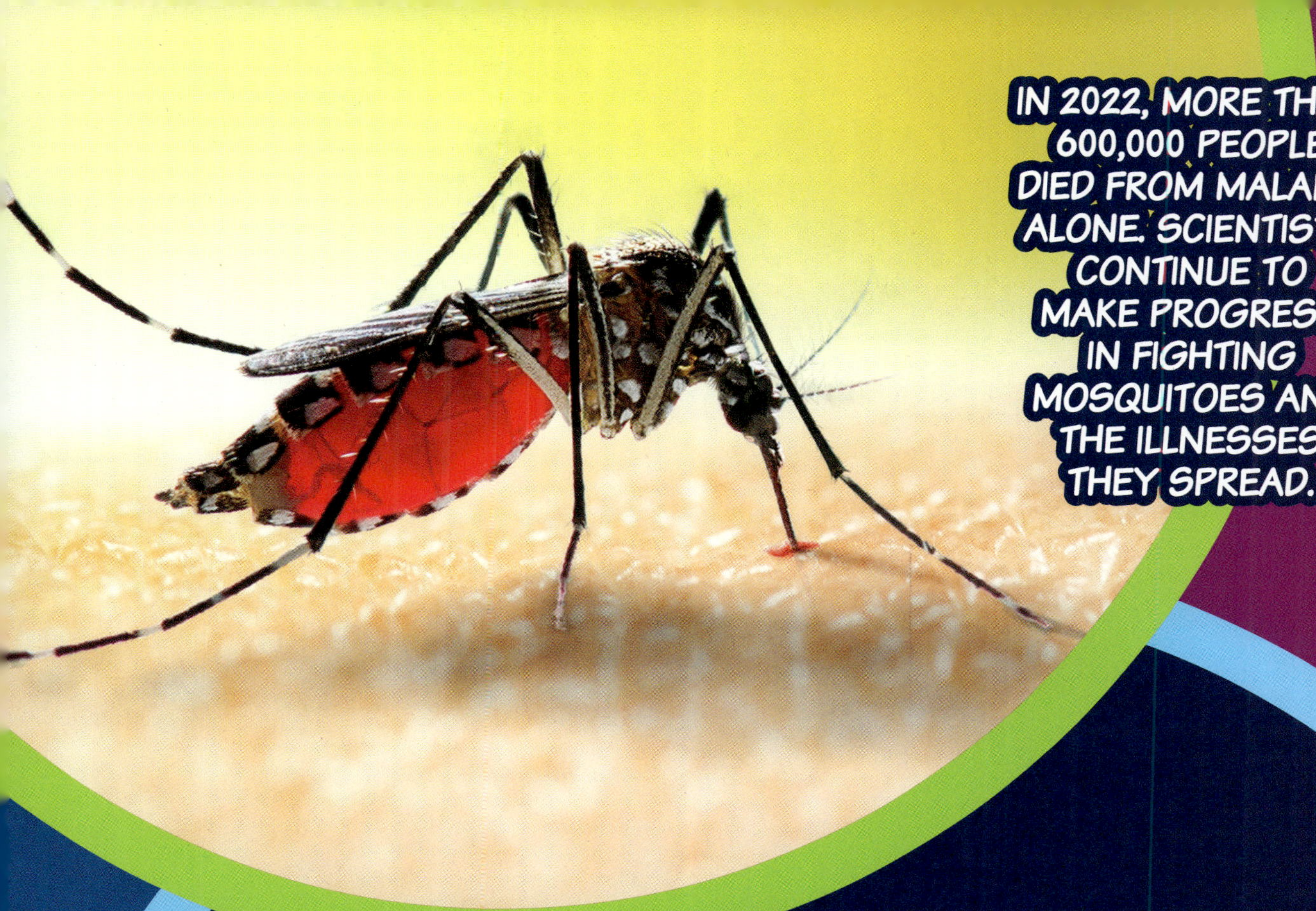

Mosquitoes are responsible for the deaths of hundreds of thousands of people every year. Some spread germs and **parasites** when they bite people and animals. Their bites can result in yellow fever, Zika fever, malaria, and other deadly illnesses.

THAT'S A LOT OF BUGS!

Scientists estimate that there 100,000 species of arachnids, some of which haven't been discovered yet! These include scorpions, mites, ticks, and daddy longlegs. Spiders make up the largest portion of arachnids with about 40,000 species. Do you think there are more or fewer species of insects?

Scientists have identified about 925,000 species of insects, but they believe there are many, many more to discover—perhaps millions! Beetles are by far the most common animal in the world. They make up 25 percent of all known life on Earth!

WHO'S HUNGRY?

It seems that with so many bugs in the world, scientists must have found uses for them. What about as food? Plenty of animals eat bugs, so why not people?

People all over the world have long used both insects and arachnids as food. It may sound gross to some, but to others it's a normal part of life. Scientists have identified several benefits to eating insects and arachnids. First, as we've said, there's a huge supply! Second, bugs have many **nutrients** people need to live.

A WORLD OF BUGS

Bugs will always be creepy to some, but they are a truly fascinating topic to explore. Getting to know insects and arachnids is just the start! Studying tiny, some not so tiny, creatures has led to some of the greatest discoveries in science, such as **vaccines**.

What else would you like to know about insects and arachnids? Keep in mind that scientists believe there could be millions of them on the planet, many of which have yet to be discovered. Keep exploring the world and keep learning more!

IN 2018, A NEW SPECIES OF TARANTULA WAS OBSERVED IN SOUTHEASTERN ARIZONA. IT HAS AN ORANGISH ABDOMEN (BACK PART) SIMILAR TO THIS TARANTULA.

GLOSSARY

arthropod: A group of animals that lack a backbone, have a segmented body, jointed limbs, and a hard outer shell.

explore: To search an area in a careful way to learn more about it.

inject: To force something into the body using a needle or sharp teeth.

magnifying lens: A shaped piece of glass that makes small things look bigger.

microscope: A tool that uses lenses and mirrors to make very small things look bigger.

nutrient: Something taken in by a plant or animal that helps it grow and stay healthy.

parasite: A living thing that lives in, on, or with another living thing and causes it harm.

vaccine: A drug made of the germs that cause illnesses that help the body fight the illness.

venom: A type of poison made by an animal and passed to another animal by a sting or bite.

wingspan: The distance between the tips of a pair of outstretched wings.

FOR MORE INFORMATION

BOOKS

Humphrey, Natalie. *Giant Wetas.* Buffalo, NY: Gareth Stevens, 2024.

Mooney, Carla. *Insects and Arachnids.* Minneapolis, MN: Adbo, 2021.

Toney, Krystal Monique. *The Fascinating Bug Book for Kids: 500 Startling Facts!* Emeryville, CA: Rockridge Press, 2022.

WEBSITES

Insects and Arthropods
easyscienceforkids.com/all-about-insects
Learn more about insects, arachnids, and other arthropods at this detailed website.

INDEX